SOUTHERN STEAM

About the Photographer

R.J. (Ron) Buckley was born in 1917 and, after the family moved to a house overlooking Spring Road station, near Tyseley, in 1926, his interest in railways grew. After joining the Birmingham Locomotive Club in 1932, he made frequent trips with them throughout the country, also accompanying W.A. Camwell on his many branch line tours. From 1936 until 1939 these club tours included visits to Southern Railway territory where he photographed many examples of the pre-grouping locomotive classes still working with that railway. In 1934 he joined the London, Midland, Scottish Railway (LM&SR) as a wages clerk at Lawley Street goods station, Birmingham, and after the declaration of war in September 1939 was called up and joined the Royal Engineers, being posted to the No. 4 Dock Operating Unit. Serving briefly in Norway during March 1940, by May that year he was with a special party supplying stores for the returning troops at Dunkirk. Ron was evacuated from Dunkirk on the *Maid of Orleans*, an ex-Southern Railway cross-Channel ferry and by 1941 he was in Egypt with his unit supporting the 8th Army in its advance from El Alamein, finally reaching Tripoli. In 1944, his unit was in Alexandria before returning to Britain in 1945, where it was demobbed in May 1946. His employment continued with the LM&SR and was based at Kings Heath in Birmingham and then at Derby in the British Railways Divisional Managers Office from 1948. He was a spectator of the continual stream of locomotives passing through the works and photographed many of the new British Railways 'Standard' locomotives constructed there. Further visits to the south of England from 1946 to 1949 and throughout the 1950s and '60s gave him the opportunity to witness the changes that had taken place in locomotive power in that region. Married in 1948 to Joyce, the daughter of a London & North Eastern Railway (LNER) locomotive driver, Ron later retired in 1977 after over forty-two years' service with the railways. He and his wife currently live in Staffordshire.

Dedication

This volume of photographs is dedicated to the memory of many former members of the Birmingham Locomotive Club, founded in 1930. Ron joined in 1932, and members accompanied him on the many shed visits and tours that the club organised. It also remembers the few current members of the club who continue to meet and follow the guiding principles set out by the first committee members.

SOUTHERN STEAM

THE RAILWAY PHOTOGRAPHS OF R.J. (RON) BUCKLEY

COMPILED BY BRIAN J. DICKSON

The History Press

First published 2016

The History Press
The Mill, Brimscombe Port
Stroud, Gloucestershire, GL5 2QG
www.thehistorypress.co.uk

British Library Cataloguing in Publication Data.
A catalogue record for this book is available from the British Library.

ISBN 978 0 7509 6613 9

Typesetting and origination by The History Press
Printed and bound in Malta by Melita Press

Front cover: Friday 22 August 1947. Designed by William Adams and introduced during 1882, the Class 415 4-4-2 tanks for the L&SWR were primarily utilised on its London suburban services. A total of seventy-one examples were constructed between 1882 and 1885, coming from four manufacturers: Beyer, Peacock & Co., Robert Stephenson & Co., Dübs & Co. and Neilson & Co. No. 3488 was a Neilson & Co. product of 1885 that would find her way into the preservation scene after withdrawal in 1961.

Back cover: Sunday 14 October 1956. Bearing a 73B shed code, ex-SR Class V 'Schools' 4-4-0 No. 30928 Stowe is seen at her home shed, Bricklayers Arms. Entering service from Eastleigh Works in 1934, she would be withdrawn in 1962 and purchased during 1964 by the Beaulieu Motor Museum for display there together with three Pullman cars. She would be moved to the East Somerset Railway in 1973 and on to the Bluebell Railway during 1980, who steamed her in 1981. She is currently in the care of the Maunsell Locomotive Society and is undergoing a complete overhaul at the Bluebell Railway.

Half-title page: Sunday 14 October 1956. The 6ft 7in diameter driving wheels of ex-SR Class N15 4-6-0 No. 30782 *Sir Pelleas* dominate this photograph of the locomotive at Nine Elms shed. An example of the class constructed during the Richard Maunsell period, she entered service from the North British Locomotive Company (NBL) in Glasgow during 1925 and would be withdrawn in 1959: a relatively short working life of only thirty-four years.

Title page: Friday 22 August 1947. At Ottery St Mary station, ex-L&SWR Class M7 0-4-4 tank No. 133 is seen departing with a train for Exmouth. Constructed at Nine Elms Works in 1903, she would be withdrawn from service during 1964.

INTRODUCTION

The creation of the Southern Railway (SR) at the grouping in 1923 brought together such disparate railway operations as the South Eastern & Chatham Railway (SE&CR), the London, Brighton & South Coast Railway (LB&SCR) and the London & South Western Railway (L&SWR). It incorporated the electrified commuter services of the LB&SCR from Victoria and London Bridge, which commenced operation as early as 1909 and reached Coulsdon on the Brighton line, together with the L&SWR electrified lines from Waterloo, which had opened during 1915 and 1916, extending as far as Claygate on the Guildford line. Prior to the electrification of these lines, both railways relied on a series of tank locomotives to handle the ever increasing demands of these commuter services. The LB&SCR 'Terrier' 0-6-0 tanks of William Stroudley design gave way to Robert Billinton's series of powerful 0-6-2 tanks in the shape of Class E3 then E4 followed by classes E5 and E6. Earle Marsh continued with tank locomotive design but in smaller quantities, with a series of classes in the 4-4-2 wheel arrangement. The SE&CR, having no electrified lines and a smaller area of commuter services, saw the beautifully balanced lines of the James Stirling Class Q 0-4-4 tanks followed by Harry Wainwright's similarly well-balanced looks of his Class H 0-4-4 tanks on these services. The L&SWR commuter services from Waterloo had been well served by the William Adams-designed Class 415 4-4-2 tanks and a dozen or so of his smaller Class O2 0-4-4 tanks that had also been allocated to this work. Adams' successor, Dugald Drummond, produced one of the heaviest and most powerful tank locomotives to handle this South London commuter traffic in the shape of his Class M7 0-4-4 tanks. So successful were these that many examples continued working throughout branches on the old L&SWR system well into the 1960s.

The compact nature of the LB&SCR operations servicing the towns and cities to the Sussex coast, and its services on the joint access line with the L&SWR into Portsmouth, saw Class B2 and B4 4-4-0 express passenger locomotives, designed by Robert Billinton, introduced from 1895, many of which were later rebuilt in more powerful forms. These were followed by the well-balanced and pleasing looks of the two Atlantic classes, H1 and H2, designed by Earle Marsh, that appeared from 1905, and, from Lawson Billinton in 1913, the very modern-looking and functional design of his Class K 2-6-0 locomotives. These proved to be so reliable that the whole class survived until withdrawal at the end of 1962.

The SE&CR had been formed with the merger of the London Chatham & Dover Railway (LC&DR) and the South Eastern Railway (SER) under a joint management committee in January 1899 and its sphere of influence incorporated the whole of Kent, with its monopoly of the Channel ports traffic and into the eastern part of Sussex reaching Hastings and Bexhill. It also went as far as Guildford and Reading through its purchase of the Reading, Guildford & Reigate Railway in 1852. Express passenger locomotives were mostly of the 4-4-0 wheel arrangement with the graceful James Stirling-designed classes from the SER sporting 7ft driving wheels. Under SE&CR management, Harry Wainwright produced a series of sure-footed reliable 4-4-0s during his tenure of office with many examples of his classes D, E and L surviving into the 1950s and '60s.

The L&SWR, in comparison with the LB&SCR and the SE&CR, had a much larger operational territory extending from Waterloo to Portsmouth and Bournemouth. It continued through Dorset and Devon to reach Exeter and Barnstaple and on into Cornwall reaching Bude and Padstow. For express passenger trains the L&SWR again relied primarily on the 4-4-0 wheel arrangement with both William Adams and Dugald Drummond producing graceful-looking, free-steaming locomotives. Adams produced eight classes and Drummond seven classes of this type, totalling over 300 examples; all the Drummond locomotives were constructed at their Nine Elms Works. Drummond attempted, unsuccessfully, to design locomotives utilising the 4-6-0 wheel arrangement and it was not until after Robert Urie arrived at the railway in 1912 that three very successful classes of 4-6-0 were produced. Class N15, which eventually formed part of the 'King Arthur' series, followed Class H15, and finally Class S15 was introduced in 1920. So successful were these three classes that, after the grouping in 1923, Richard Maunsell continued construction, with the last of the Class S15s appearing during 1936.

Richard Maunsell's years at the SE&CR (1913–22) saw the introduction of the 2-6-0 wheel arrangement to that railway with his Class N of 1917, of which twelve examples were constructed up to 1922, with a further sixty-eight appearing during his period at the Southern Railway. A design for a three-cylinder version of this class was prepared during his SE&CR days, with all six examples of this Class N1 entering service after the grouping. The years Maunsell spent as chief mechanical engineer (CME) with the Southern Railway, from 1923 until his retirement in 1937, saw a steady output of new classes, including the four-cylinder 4-6-0 'Lord Nelsons' that proved to be poor performers. Another highly successful class of 2-6-0s, the Us, appeared during 1928, followed by a three-cylinder version, the U1s. In 1930, what has been described as Maunsell's masterpiece design, the Class V or 'Schools' three-cylinder 4-4-0, was introduced. Numbering forty examples constructed at Eastleigh Works and entering service between 1930 and 1935, they were all withdrawn in 1961–62.

Oliver Bulleid became the Southern Railway CME during 1937 and in 1938 was given authority to commence with design work for a new main-line express passenger locomotive, with the first of these revolutionary locomotives entering service in June 1941. The 'Merchant Navy' Class 4-6-2s utilised an all-steel welded firebox incorporating thermic syphons, a boiler pressure of 280psi and a totally enclosed chain-driven valve gear for all three cylinders. Fitted with a steam reverser and steam-operated firebox doors, the omission of dampers on the ash pan to control primary air through the fire grate was an unusual innovation. All this was enclosed in a rather unconventional-looking air-smoothed casing, with a total of thirty examples constructed at Eastleigh Works between 1941 and 1949, all being rebuilt in a conventional form between 1956 and 1959.

Four years after the introduction of the 'Merchant Navy' class, in May 1945, a lighter version, the 'West Country' and 'Battle of Britain' classes, started to roll out of Brighton Works to much fanfare by the publicity department of the Southern Railway. At approximately 10 tons lighter than the 'Merchant Navy' class, their route availability was much increased, which meant that they could work to the most westerly outposts of the railway, Padstow and Ilfracombe. A total of 110 examples were built, all coming from Brighton Works except for six examples that were constructed at Eastleigh Works during 1949–50. A total of sixty examples were rebuilt in a conventional form between 1957 and 1961.

These Pacific locomotives were not altogether a success operationally, with their chain-driven valve gears in their 40-gallon oil baths and their coal consumption causing concern. The decision was made during 1955 to commence the rebuilding of the classes in a more conventional form, replacing the unique Bulleid valve gear with a Walschaerts valve gear operated by a screw reverser; a new superheater header was fitted, the working boiler pressure was reduced to 250psi and a new ash pan was fitted with dampers to assist in the control of primary air. Finally the air-smoothed cladding was removed and smoke deflectors were fitted. The result was a locomotive with much improved performance and easier access for maintenance.

Ron Buckley's photographs show the changing locomotive scene taking place throughout the Southern Railway and its successor, the Southern Region of British Railways, illustrating from the late 1930s those pre-grouping classes that were still working. These include the work of such well-known designers as William Adams, Dugald Drummond and Robert Urie of the L&SWR; William Stroudley, the Billintons and Earle Marsh of the LB&SCR; and Harry Wainwright and Richard Maunsell of the SE&CR. Ron's later photographs, from 1946 onwards, continue to show some of those pre-grouping locomotives and additionally portray the newer Oliver Bulleid-designed Pacifics introduced during 1941. Visits to the Isle of Wight during the 1940s, '50s and '60s enabled him to capture images of almost the entire fleet of locomotives working there during those three decades.

Sunday 25 October 1936. Regarded as one of Richard Maunsell's outstanding designs, the 'Schools' were free-steaming, reliable locomotives, very much liked by their crews. Barely two years into its service life, SR Class V 'Schools' 4-4-0 No. 926 *Repton* is seen in the yard at Fratton shed, Portsmouth. Constructed at Eastleigh Works and entering service during May 1934, she would be withdrawn in December 1962 and purchased by a museum in the USA, arriving there during 1967. A move to Canada followed in 1974 with a return to the USA during 1979. It was April 1989 before she arrived back in the UK, to be based at the North Yorkshire Moors Railway where she is currently undergoing a major overhaul.

Sunday 25 October 1936. Parked in the yard at Fratton shed is ex-LB&SCR Class A1X 'Terrier' 0-6-0 tank No. B644. Designed by William Stroudley and constructed at Brighton Works as a Class A1 locomotive, she entered service during 1877 numbered 44 and named *Fulham*. She would be rebuilt and modified in 1912 to re-enter service as a Class A1X locomotive. Destined to give seventy-four years of service, she would eventually be withdrawn, having been numbered 32644 in 1951.

Sunday 25 October 1936. Another 'Terrier' 0-6-0 tank seen here at Fratton is No. 2653, originally No. 53 *Ashtead*. Constructed as part of the class of fifty A1 locomotives at Brighton Works for the LB&SCR, she entered service in 1875. Rebuilt in Class A1X form during 1912, she was sold to the Weston, Clevedon & Portishead Railway during 1937 and was scrapped in 1948.

Sunday 26 September 1937. Another successful William Stroudley design for the LB&SCR was the Class E1 0-6-0 tank, of which a total of seventy-nine were constructed over a period of seventeen years at its Brighton Works. Seen here at Bricklayers Arms shed is No. 2164, which entered service during 1891, numbered 164 and named *Spithead*; she was withdrawn in 1948.

Sunday 26 September 1937. James Stirling introduced the versatile Class O 0-6-0 to the SER during 1878, with building continuing for twenty-one years until 1899; a total of 122 examples entered service. Ashford Works built fifty-seven of the class, the remainder coming from Sharp, Stewart & Co. Many examples were rebuilt during the Wainwright years to become Class O1 and gave an improved performance. The example seen here at Bricklayers Arms shed, No. 1093, was constructed at Ashford Works during 1897 and rebuilt as Class O1 in 1916. She would be withdrawn from service in 1951.

Sunday 26 September 1937. In the yard at Bricklayers Arms shed is SR Class L1 4-4-0 No. 1758. Richard Maunsell modified the earlier Wainwright design of Class L locomotives to improve performance and all fifteen examples of the L1 were constructed by the North British Locomotive Company (NBL) and delivered during 1926. Initially introduced for the London to Folkestone express passenger traffic, they migrated to other sections after electrification of the Kent coast lines. The locomotive seen here would become No. 31758 with British Railways and be withdrawn during 1959.

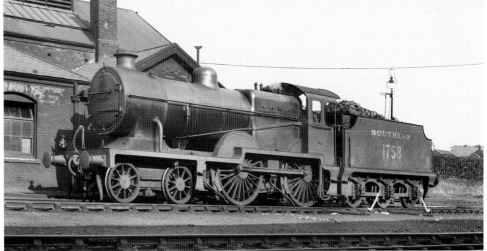

Sunday 26 September 1937. Seen here at New Cross Gate shed is Robert Billinton-designed Class E6 0-6-2 tank No. 2417. Built at the Brighton Works of the LB&SCR during 1905, she would become No. 32417 with British Railways and be withdrawn in 1962.

Sunday 26 September 1937. New Cross Gate shed is host to ex-LB&SCR Class E1 0-6-0 tank No. 2142. Designed for shunting and light goods traffic, Class E1 locomotives could also undertake some passenger turns as a number were fitted with the Westinghouse brake. Constructed at Brighton Works in 1879, numbered 142 and named *Toulon*, she was destined to be numbered 31242 with British Railways and be withdrawn after seventy-one years of service during 1950.

Sunday 26 September 1937. In beautifully clean condition at New Cross Gate shed, ex-LB&SCR Class I3 4-4-2 tank No. 2075 is glistening in the late summer sunshine. Designed by Douglas Earle Marsh specifically for the London to Brighton express passenger services, the class of twenty-seven examples was constructed at Brighton Works between 1907 and 1913. No. 2075 was a 1910-built example that acquired superheating during 1925. Replaced by larger, more powerful locomotives, the class migrated to duties from sheds such as Salisbury and Dover. She would be numbered 32075 by British Railways and withdrawn during 1951.

Sunday 31 October 1937. Standing in the yard at Ashford shed is ex-SE&CR Class E 4-4-0 No. 1176. Designed by Harry Wainwright as an improved version of his Class D locomotives utilising a Belpaire firebox as opposed to the round-topped variety, all twenty-six examples were constructed at Ashford Works between 1906 and 1909. Entering service during 1907, she would be withdrawn in 1951 numbered 31176.

Sunday 31 October 1937. The well-balanced, graceful lines of Wainwright-designed ex-SE&CR Class D 4-4-0s from 1901 are seen here with No. 1743 standing in the yard at Ashford shed. Constructed by Robert Stephenson & Co. during 1903, she would be rebuilt at Ashford Works as a Class D1 locomotive in 1927. One of the longest serving members of the class, she would survive fifty-seven years to be withdrawn in 1960 numbered 31743.

Sunday 19 June 1938. William Adams' career with the L&SWR saw him produce a series of express passenger 4-4-0s, three of which utilised driving wheels of 7ft 1in diameter. His final design, Class X6 introduced in 1895, used 6ft 7in driving wheels and the ten locomotives of the class were constructed at Nine Elms Works. The result was a particularly graceful-looking locomotive as seen here. *Right:* Parked in the yard at Basingstoke shed is No. 666, the last of the class to be constructed and still retaining its original Adams-style safety valves but fitted with a Drummond-type chimney. She would be withdrawn during 1943. *Below right:* Seen in Eastleigh shed yard is classmate No. 658, sporting a Drummond replacement boiler with dome-mounted safety valves and the Drummond chimney. She would be the last of the class to be withdrawn in 1946.

Sunday 19 June 1938. The demand for a more powerful locomotive to work the LB&SCR suburban traffic around London led William Stroudley to design the Class D1 0-4-2 tank as a replacement for his earlier 'Terrier' design. Introduced during 1873, construction continued until 1887 with a total of 125 examples entering service, thirty-five of which had been built by Neilson & Co. in Glasgow. Seen here in Eastleigh Works yard is No. 2261, formerly LB&SCR No. 261 *Wigmore*, which was officially withdrawn the month prior to this photograph. She was one of the Neilson & Co. constructed examples entering service in 1882.

Sunday 19 June 1938. This well-proportioned locomotive standing in the yard at Eastleigh shed is Robert Urie-designed Class H15 4-6-0 No. 521. Originally designed as a mixed-traffic locomotive for the L&SWR and introduced during 1914, the example seen here was constructed at Eastleigh Works in 1924. Fitted with a superheated boiler, this was a strong and powerful class of locomotive. The example seen here was withdrawn in 1961.

Sunday 19 June 1938. Designed specifically for heavy shunting duties at the Feltham Marshalling Yard and some cross-London goods traffic, the four locomotives that comprised the G16 Class of 4-8-0 tanks were massive in comparison to any other L&SWR locomotives. Designed by Robert Urie and constructed at Eastleigh Works during 1921, all four examples spent the majority of their working lives at Feltham. No. 492 has the power classification letter 'A' painted close to the front buffer beam and is seen here at Eastleigh shed in sparkling condition, probably after an overhaul. She would be withdrawn in 1959 numbered 30492.

Sunday 19 June 1938. Dugald Drummond had produced many successful designs for both the North British and Caledonian railways before arriving as locomotive superintendent at the L&SWR in 1895. From 1898 on, he produced a series of 4-4-0 designs of which the Class L11 was probably the most graceful looking. Seen here at Eastleigh shed, in ex-works condition and bearing an 'F' power classification, is No. 155 which is attached to a 'water cart' style of tender. Constructed at Nine Elms Works during 1903, she would be withdrawn in 1951.

Sunday 19 June 1938. Robert Urie's third design for the L&SWR would be another 4-6-0 locomotive based on his successful H15 and N15 classes using the same wheel arrangement. Primarily designed to handle goods traffic, the first batch of Class S15 appeared during 1920 from Eastleigh Works. The example seen here at Eastleigh shed is No. 831, constructed during 1927, and is looking in ex-works condition. Bearing the SR power classification code letter 'A', and with one smoke deflector yet to be re-fitted, she became No. 30831 with British Railways and would be withdrawn during 1963.

Sunday 10 July 1938. Parked in the yard at Exmouth Junction shed is ex-L&SWR Class T9 4-4-0 No. 289. Another of the successful designs from Dugald Drummond, this example was one of the Nine Elms Works batch of 1900. Originally paired with a six-wheeled tender, this would later be exchanged for the eight-wheeled 'water cart' variety seen here. Finally numbered 30289, she would be withdrawn from service during 1959.

Sunday 10 July 1938. Also seen in the yard at Exmouth Junction shed is ex-L&SWR Class M7 0-4-4 tank No. 24. Designed in response to the demand for a more powerful locomotive to handle London suburban traffic, Dugald Drummond drew on his previous experience in designing the Class 171 0-4-4 tanks for the Caledonian Railway. Introduced in 1897, the M7s were large, handsome, well-proportioned locomotives and proved themselves to be reliable, sure footed and free steaming, with the result that 105 examples were constructed at both Nine Elms and Eastleigh works over a period fourteen years. The example seen here is one of the 1899 batch from Nine Elms Works that would survive sixty-four years to be withdrawn during 1963.

Sunday 10 July 1938. William Stroudley introduced his Class E1 0-6-0 tanks to the LB&SCR during 1874 and a total of seventy-nine were constructed over a period of seventeen years. Ten of the class were rebuilt by the Southern Railway between 1927 and 1929 to become 0-6-2 tanks classified E1/R. Intended for work in the North Devon and Cornwall branches, several finished their working lives undertaking banking duties between Exeter St Davids and Central stations. Seen here at Exmouth Junction shed is No. 2695, originally entering service as Class E1 No. 103 *Normandy* in 1876; she would be rebuilt during 1928 as Class E1/R and be withdrawn after a total of eighty-one years of service in 1957.

Sunday 10 July 1938. Following the Maunsell experimental work on converting a three-cylinder Class K1 'River' 2-6-4 tank to a three-cylinder 2-6-0 tender locomotive during 1928, a batch of twenty new locomotives to be classified U1 was built at Eastleigh Works in 1931. Designed to work passenger traffic, they proved themselves to be very reliable and successful locomotives. No. 1896 is seen here in the yard at Exmouth Junction shed, destined to be numbered 31896, she would be withdrawn from service during 1962.

Sunday 10 July 1938. The locomotive seen here was something of a star, having been exhibited by Dübs & Co. at the Glasgow Exhibition immediately after her construction during 1901. Ex-L&SWR Class T9 4-4-0 No. 733 was sold to the railway company after closure of the exhibition and is seen here, light engine, at Exeter Central station. She would be withdrawn from service during 1952.

Sunday 12 March 1939. Stewarts Lane shed is playing host to ex-LB&SCR Class E5 0-6-2 tank No. 2583. Designed by Robert Billinton as a larger version of his Class E4 locomotives, they were primarily used on the outer suburban routes. Constructed at Brighton Works during 1903 and entering service as No. 583 *Wandcombe*, she would become No. 32583 with British Railways and be withdrawn in 1956.

Sunday 4 June 1939. The turntable at Reading shed is being used to turn SR Class U 2-6-0 No. 1800. Entering service from Ashford Works during 1928, this locomotive was one of the twenty examples of the class built using the redundant 'River' K Class 2-6-4 tanks as their foundation. She would be withdrawn in 1965.

Sunday 31 March 1946. Another class of locomotive originating as a rebuild from earlier designs were the N15X 4-6-0s appearing during the Richard Maunsell period between 1934 and 1936. First introduced in 1914 as Lawson Billinton-designed and Brighton Works-built Class L 4-6-4 tanks, the seven members of the class were utilised on the London to Brighton line. Becoming redundant on the electrification of that line in 1933, they were transferred to Eastbourne to work the London traffic from there. Finally out of service during 1934–35, they were used as the basis for the N15X locomotives, all seven being rebuilt between 1934 and 1936 at Eastleigh Works. Seen here standing in the yard at Eastleigh shed is No. 2328 *Hackworth*, which would become No. 32328 and be withdrawn during 1955.

Sunday 31 March 1946. Designed by William Adams as shunting and station pilot locomotives, the Class B4 0-4-0 tanks, introduced in 1891, quickly migrated to work as dock tanks, primarily at Southampton, which the L&SWR owned. No. 97 *Brittany*, seen here in Southampton Docks, was constructed at Nine Elms Works during 1893 and would be withdrawn from service in 1949.

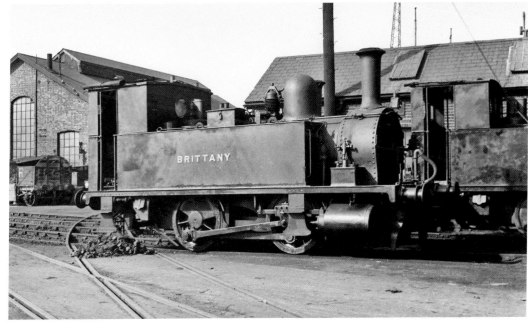

Saturday 16 August 1947. During 1932 and 1933, four Stroudley-designed Class E1 0-6-0 tanks were transferred to the Isle of Wight, primarily to work the goods traffic on the island. On this bright summer's day at Newport shed we see No. 2 *Yarmouth*. Constructed at Brighton Works during 1880 and entering service as No. 152 *Hungary*, she would be withdrawn after seventy-six years of service during 1956.

Saturday 16 August 1947. Seen here entering Newport station at the head of a passenger train from Cowes is ex-L&SWR Class O2 0-4-4 tank No. 28 *Ashey*. An 1890-built example of the class, originally numbered 186, she would be transferred to the island during 1926 and acquire her new number and name. She would be withdrawn at the closure of the island's railways in December 1966.

Saturday 16 August 1947. Looking splendid in Southern Railway livery with 'sunshine lettering' is ex-LB&SCR Class E1 0-6-0 tank No. 4 *Wroxall*. Built at Brighton Works during 1878 and numbered 131 *Gournay*, she would be transferred to the Isle of Wight in 1933 and be numbered and named as seen here. Withdrawn during 1960, she would have given a total of eighty-two years of service.

Saturday 16 August 1947. Ex-LB&SCR Class E1 0-6-0 tank No. 3 *Ryde* is seen entering Newport station with a train of loaded coal wagons from Medina Wharf. Constructed at Brighton Works in 1881 and entering service as No. 154 *Madrid*, she would be transferred to the island during 1932 and receive her new number and name. Withdrawal would take place in 1959.

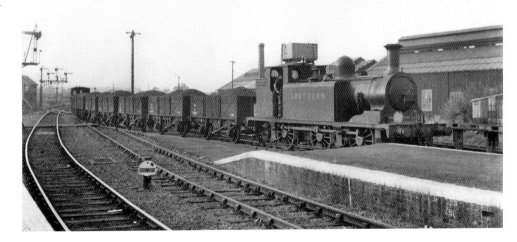

Sunday 17 August 1947. Parked adjacent to the shed at Salisbury is Urie-designed ex-L&SWR Class S15 4-6-0 No. 502. Built at Eastleigh Works during 1920 as one of the first batch of this successful class, she would be based at Feltham shed for the bulk of her working career and be withdrawn numbered 30502 during 1962.

Tuesday 19 August 1947. Between 1898 and 1912, Dugald Drummond produced no less than seven classes of 4-4-0 tender locomotives for the L&SWR, the third of which comprised the forty examples of the Class K10 that were constructed at Nine Elms Works during 1901 and 1902. Seen here in grimy post-war condition standing adjacent to Bude shed, is No. 329, a late 1901-built example that still bears the graceful Drummond lines; she would be withdrawn during 1950. Seen in the background, the shed was an original L&SWR construction dating from 1898.

Tuesday 19 August 1947. At Barnstaple shed, the fireman on SR Class 'West Country' 4-6-2 No. 21C101 *Exeter* is happy to pose for the camera while checking that the headboards for the *Devon Belle* are fitted correctly. The first of Oliver Bulleid's Light Pacific design to enter service in June 1945 from Brighton Works, she is seen here with the longer smoke deflectors that were fitted at a later date. Late in 1957 she would emerge from Eastleigh Works in rebuilt form without the air-smooth casing and, after only twenty-two years of service, be withdrawn numbered 34001 in 1967.

Friday 22 August 1947. Seen here at Axminster station waiting to depart with the 4.40 p.m. train to Lyme Regis is ex-L&SWR Class 415 4-4-2 tank No. 3488. Designed by William Adams and entering service during 1885 from the Glasgow Works of Neilson & Co., this locomotive would be sold into industrial use in 1917 and subsequently find her way to work on the East Kent Railway, finally arriving into Southern Railway ownership in 1946. She would be withdrawn numbered 30583 by British Railways during 1961 and sold to the Bluebell Railway where she is currently on static display.

Saturday 20 March 1948. Standing in the yard at Hither Green shed is Harry Wainwright-designed ex-SE&CR Class C 0-6-0 No. 1460. The first of nine examples of the class, totalling 109 locomotives, to appear from the ex-LC&DR Works at Longhedge, she would enter service during 1902 and be withdrawn after forty-seven years of service in 1949.

Saturday 20 March 1948. Designed to handle inter-regional goods traffic in the London area, the Richard Maunsell-designed three-cylinder SR Class W 2-6-4 tanks had their design origins in the earlier three-cylinder 'River' Class tanks. Seen here at Hither Green shed is Ashford Works-constructed No. 1921, which entered service during 1935. This powerful class of locomotive proved itself a reliable and efficient workhorse for both the Southern Railway and British Railways. No. 1921 would be withdrawn in 1963 numbered 31921 by British Railways.

Saturday 20 March 1948. Constructed at Brighton Works between 1913 and 1921, the seventeen examples of Lawson Billinton's Class K 2-6-0 for the LB&SCR were largely based at both Brighton and Three Bridges sheds throughout their working lives. Seen at New Cross Gate shed is No. 2351, one of the 1921-built examples of the class. With clean, workman-like lines, they were constructed using Belpaire fireboxes and superheated boilers, and proved themselves to be very reliable and surefooted. The whole class of seventeen was withdrawn from service during November and December 1962.

Saturday 20 March 1948. Glistening in the late winter sunshine at New Cross Gate shed is ex-LB&SCR Class H2 4-4-2 No. 2426 *St Albans Head*. Designed by Douglas Earle Marsh but introduced by Lawson Billinton in 1911, after the early departure of Marsh due to ill health, the six examples of the class were constructed at Brighton Works using superheated boilers, which gave them a much-improved performance in comparison to the earlier Class H1 Atlantics. All six locomotives acquired their names during Southern Railway days and No. 2426, the last to be constructed in 1912, would be withdrawn in 1956 numbered 32426.

Saturday 20 March 1948. Sporting her new owner's name in Southern-style lettering and numbered S2455 is ex-LB&SCR Class E3 0-6-2 tank. A product of Brighton Works during 1895, she was originally numbered 455 and named *Brockhurst*. She was destined to bear the British Railways number 32455 and be withdrawn in 1958.

Saturday 20 March 1948. Still bearing the word 'Southern' on her tank side is ex-LB&SCR Class E4X 0-6-2 tank No. 2489. Constructed at Brighton Works originally to the design of Robert Billinton as a Class E4 locomotive, she would be rebuilt with a larger boiler during 1909 by Earle Marsh and classified E4X. Entering service in 1899, numbered 489 by the LB&SCR and named *Boxgrove*, she would finally be withdrawn during 1955.

Saturday 20 March 1948. Seen here at Feltham shed is ex-L&SWR Class N15 'King Arthur' 4-6-0 No. 742 *Camelot*. Designed by Robert Urie for express passenger duties, she was constructed at Eastleigh Works during 1919 and would be based at Bournemouth for a large part of her working life. Originally unnamed, it was after the Southern Railway built a further fifty-four examples during the mid-1920s that they decided to name the locomotives in the class – the original twenty examples acquired names of places and characters associated with the King Arthur legends. The locomotive seen here was rebuilt with a superheated boiler during 1930 and withdrawn from service in 1957.

Wednesday 15 June 1949. Seen here at Wadebridge are the three veteran 2-4-0 Well tank locomotives originally designed by Joseph Beattie and constructed by Beyer, Peacock & Co. They were rebuilt over a period of years by his successors at the L&SWR and finally designated Class 0298. Allocated to work the china clay traffic between Wadebridge and Wenford Bridge, all three survived until late in 1962, when they were replaced by more modern ex-GWR locomotives. *Left:* No. 30585, constructed in 1874, would be numbered 314 by the L&SWR and she would later receive a '0' prefix as part of their duplicate list. Fortunately she would reach the preservation scene and is currently in the care of the Buckinghamshire Railway Centre at their Quainton Road museum. *Right:* No. 30586 was built in 1875 and numbered 329 by the L&SWR, again to be placed on their duplicate list and bear the '0' prefix. She would unfortunately not enter the preservation scene after withdrawal. *Below right:* No. 30587 was constructed during 1874 and numbered 298 and later 0298 in the L&SWR duplicate locomotive list. After withdrawal she was retained for the National Collection and is currently in the care of the Bodmin & Wenford Railway, which is appropriate after working in that area for so many years. Note the different shaping to the splashers and the style of lettering and numbering on the three locomotives.

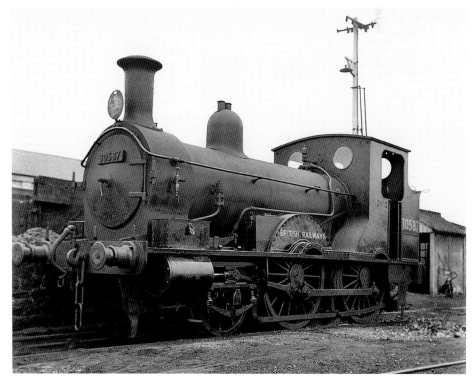

Saturday 15 March 1952. Parked outside the shed at Newhaven is ex-LB&SCR Class A1X 'Terrier' 0-6-0 tank No. 32636. Built in 1872 as a Class A1 locomotive at Brighton Works, she was originally numbered 72 and named *Fenchurch*. She would be rebuilt as Class A1X during 1913. Sold into industrial use in 1898, she would return to Southern Railway ownership during 1927. Withdrawn in 1963, she was purchased for the Bluebell Railway and is currently not operational.

Saturday 15 March 1952. Seen here at Newhaven shed is the last-built example of ex-LB&SCR Class K 2-6-0 No. 32353. After entering service during 1921, she would be withdrawn after only forty-one years of service in 1962.

Saturday 15 March 1952. Newhaven shed is host to ex-LB&SCR Class E4 0-6-2 tank No. 32511. Designed by Robert Billinton and built at Brighton Works, she entered service during 1901, originally numbered 511 and named *Lingfield* by that railway. She would be withdrawn from service in 1956.

Saturday 15 March 1952. Diminutive Harry Wainwright-designed ex-SE&CR Class P 0-6-0 tank No. 31556 is parked in the yard at Brighton shed. Introduced during 1909, this class of eight examples was primarily constructed to replace the unsatisfactory steam railcars operated by that railway. All built at Ashford Works, the locomotive seen here was constructed in 1909 and originally bore the number 753. Withdrawn during 1961, she was sold into industrial use and finally to the Kent & East Sussex Railway during 1970 where she is currently undergoing a major overhaul.

Saturday 15 March 1952. The original design for what became the LB&SCR Class E3 0-6-2 tank was the work of William Stroudley, but with his death in 1889 work was delayed until his successor Robert Billinton made modifications to the designs. Eventually appearing in 1894 from Brighton Works, the class was well suited to mixed-traffic work and shunting duties. Seen here in Brighton shed yard is No. 32169; entering service in 1894 as No. 169 and named *Bedhampton*, she would be withdrawn during 1955.

Saturday 15 March 1952. Seen here at Brighton shed is ex-LB&SCR Class D3 0-4-4 tank No. 32368. Designed by Robert Billinton to operate semi-fast passenger trains, locomotives of this class spent most of their days working on the Kent and Sussex lines. Constructed at Brighton Works during 1892, she would be numbered 368, named *Newport* and withdrawn in 1953 after sixty-one years of service.

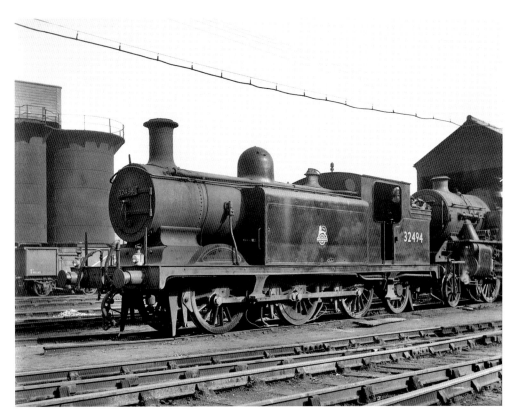

Saturday 15 March 1952. In reasonably clean condition, ex-LB&SCR Class E4 0-6-2 tank No. 32494 is parked in Brighton shed yard. One of the earlier constructed examples of 1899 of a class totalling seventy-five locomotives, she was numbered 494 and named *Woodgate*, surviving sixty-one years to be withdrawn from service during 1959.

Saturday 15 March 1952. Harry Wainwright-designed ex-SE&CR Class E1 4-4-0 No. 31507 is seen here in the yard at Brighton shed. Originally constructed at Ashford Works in 1908 as a Class E, she would be rebuilt in the form seen here by Beyer, Peacock & Co. in 1920 during the Richard Maunsell period at that railway. Originally numbered 507 by the SE&CR, she would become No. 1507 with the Southern Railway and finally No. 31507 with British Railways. She was withdrawn in 1961.

Saturday 15 March 1952. Another Wainwright-designed ex-SE&CR 4-4-0 seen here at Brighton shed is the Class D No. 31733 sporting a vague outline of both Southern and British Railways ownership on her tender. Redolent of an earlier period of locomotive design, her lines certainly have a grace that was lost in later 4-4-0 designs. Constructed by Sharp, Stewart & Co. during 1901, she avoided the rebuilding of some of the class by Maunsell, therefore retaining the round top firebox, and served for fifty-two years before being withdrawn in 1953.

Saturday 15 March 1952. Prior to the introduction of the British Railways Standard locomotives and in an effort to quickly replace the many worn out tank locomotives inherited after the war, British Railways continued the building of the Fairburn LMS Class 4 2-6-4 tanks, placing an order for forty-one examples to be constructed at Brighton Works. These started appearing during 1950 and seen here is No. 42093, a product of the works, standing in the yard at Brighton shed. She would spend the bulk of her working life based at 55F Bradford Manningham shed and be withdrawn after only sixteen years of service in 1967.

Saturday 15 March 1952. The cleaning staff at Brighton shed have done a wonderful job as ex-LB&SCR Class H2 4-4-2 No. 32422 *North Foreland* reflects the afternoon sunshine. Built at Brighton Works during 1911, she would be withdrawn in 1956.

Saturday 15 March 1952. This sturdy-looking locomotive parked at Brighton shed is ex-L&SWR Class G6 0-6-0 tank No. 30266. One of the few remaining examples of a class totalling thirty-four still to be found working at this time, she was the product of Nine Elms Works in 1894 and was numbered 266 by that railway. Designed by William Adams for yard shunting duties, the first ten examples appeared in 1894 with further batches being constructed, the last to appear in 1900. The locomotive seen here has been fitted with shunter's steps and bunker handrails, and was withdrawn during 1960 having given sixty-six years of service.

Saturday 15 March 1952. Seen here at Brighton shed is ex-LB&SCR Class E4 0-6-2 tank No. 32577. A 1903-constructed member of this class, she was originally numbered 577 and named *Blackstone*. Designed for mixed-traffic work, these locomotives were both powerful and free-steaming examples of the designer's work. She would see fifty-six years of service before being withdrawn during 1959.

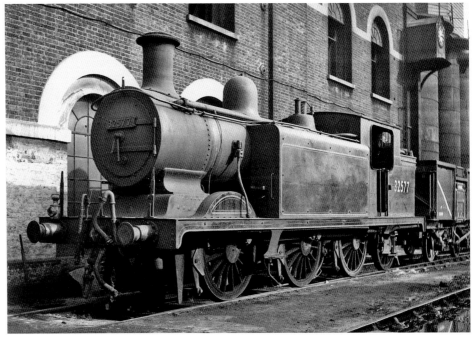

Saturday 15 March 1952. Ex-SE&CR Class D1 4-4-0 No. 31739 is standing adjacent to the ash pits at Brighton shed whilst the fireman is seen trimming the coal in the tender. Originally constructed as a Wainwright-designed Class D at Ashford Works in 1902, she would be rebuilt in the form seen here, with a Belpaire firebox, at Ashford Works during 1927. She was withdrawn from service in 1961.

Saturday 30 August 1952. In ex-works condition, ex-L&SWR Class M7 0-4-4 tank No. 30243 is seen here at Eastleigh shed. One of the earliest members of the class, a product of Nine Elms Works during 1897, she would be withdrawn from service sixty-one years later in 1958.

Saturday 30 August 1952. Still bearing the legend 'Southern' on her tender, ex-SR Class H15 4-6-0 No. 30524 has acquired her new British Railways number. The last of the class to be constructed during the Maunsell period in 1924, she would be withdrawn from service during 1961.

Saturday 30 August 1952. Ex-LB&SCR Class H2 4-4-2 No. 32424 *Beachy Head* is seen here in the yard at Eastleigh shed. Entering service from Brighton Works during 1911, she would be withdrawn forty-seven years later in 1958. In 2000 the Bluebell Railway Atlantic Group was formed to build a H2 Atlantic locomotive in the form of a re-created *Beachy Head*. A suitable boiler was available and currently a rolling chassis with cylinders and motion is almost complete.

Saturday 30 August 1952. Seen here in the yard at Eastleigh shed is an example of both the Oliver Bulleid-designed Pacific Classes. *Right:* Seen here is 'Merchant Navy' Class No. 35002 *Union Castle*. Constructed at Eastleigh Works in 1941 and numbered 21C2, she would be rebuilt by British Railways at Eastleigh Works during 1958 and withdrawn in 1964. *Below right:* For comparison, 'Battle of Britain' Class No. 34088 *213 Squadron* was a product of Brighton Works during 1948. She would be rebuilt by British Railways in 1960 and withdrawn by them in 1967. For many years based at Stewarts Lane shed (73A), she was very much a favourite of the staff there, who kept her in immaculate condition for working the 'Golden Arrow' and Royal Train duties.

Saturday 30 August 1952. Designed by Richard Maunsell specifically for heavy-shunting duties in yards such as Eastleigh, Exmouth Junction and Hither Green, the three-cylinder Class Z 0-8-0 tanks, introduced by the Southern Railway in 1929, comprised only eight examples. Seen here is No. 30950, numerically the first of the class, at Eastleigh shed. By the late 1950s all eight members of the class were working out of Exmouth Junction shed on banking duties at Exeter. The whole class would be withdrawn from service towards the end of 1962.

Saturday 30 August 1952. Seen here at Eastleigh shed is the unusual sight of a locomotive constructed by a German manufacturer for a British railway company. Designed for the SE&CR by Harry Wainwright but delivered during the Richard Maunsell period, in 1914, the last ten examples of Class L 4-4-0 arrived from Borsig of Berlin in kit form and were erected by Borsig employees at Ashford Works. Seen here at Eastleigh shed is No. 31772, she would be withdrawn during 1959.

Saturday 30 August 1952. Standing in the yard at Eastleigh shed is Class 2 2-6-2 tank
No. 41305. Constructed at Crewe Works only four months earlier, she was part of a batch
of thirty locomotives from this class allocated to sheds throughout the Southern Region as
distant as Exmouth Junction and Faversham. She would end her days based at Weymouth
shed and be withdrawn in 1965.

Saturday 30 August 1952. Standing outside the shed at Eastleigh, ex-LB&SCR Class E2 0-6-0 tank No. 32103 is in good clean condition. Designed by Lawson Billinton and constructed at Brighton Works between 1913 and 1916, the ten members of the class were found various duties, including some London suburban routes, empty coaching stock workings and local goods traffic in the London area. The locomotive seen here was a 1913-built example that would be withdrawn from service during 1962.

Saturday 30 August 1952. Dugald Drummond's final design of the 4-4-0 express passenger locomotive for the L&SWR was the Class D15, a large and powerful free steamer. Only ten examples were constructed, all of which came from Eastleigh Works in 1912. Allocated initially to work the Bournemouth expresses, they migrated to the Portsmouth line and some cross-country services later in life. Designed with saturated boilers, they were rebuilt by Urie with superheaters, which meant an extension to the smoke box and, with the change to the stovepipe chimneys, it did nothing to improve their overall looks, as can be seen here. No. 30466 is standing at Eastleigh shed a month before her official withdrawal during September 1952.

Saturday 30 August 1952. A most unorthodox design of locomotive stemmed from the need for austerity during wartime, which led to the Oliver Bulleid Class Q1 0-6-0 introduced in 1942. A total of forty examples of the class were constructed at both Ashford and Brighton Works, and they were allocated to goods traffic duties throughout most of the Southern Railway. Seen here at Eastleigh shed is No. 33020, a product of Ashford Works in 1942. She would be one of the last of the class to be withdrawn, during 1966.

Saturday 30 August 1952. This well-proportioned and balanced-looking locomotive standing in the yard at Eastleigh shed is ex-L&SWR Class O2 0-4-4 tank No. 30225. Constructed at Nine Elms Works to a design by William Adams during 1892, she would survive for seventy years before being withdrawn in 1962.

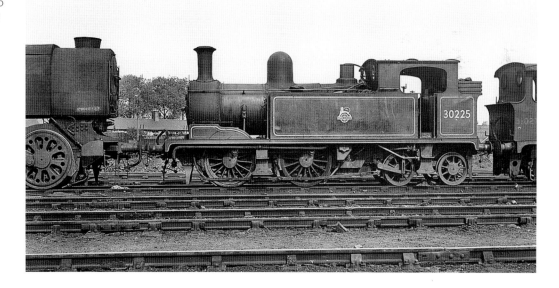

Saturday 30 August 1952. With her number and ownership lettering barely visible, ex-SE&CR Class P 0-6-0 Tank No. 31027 is seen at Eastleigh shed. Constructed at Ashford Works during 1910, she would be withdrawn in 1961 and sold to the Bluebell Railway, where she is currently based and awaiting an overhaul.

Saturday 30 August 1952. Seen here at Eastleigh shed is ex-LB&SCR Class E6
0-6-2 tank No. 32416. Built at Brighton Works in 1905 and originally numbered
416 with that railway, she would be withdrawn during 1962.

Saturday 30 August 1952. The veteran locomotive seen here at Eastleigh shed is ex-L&SWR Class 0395 0-6-0 No. 30566. Introduced in 1881 to a design by William Adams, a total of seventy examples were constructed by Neilson & Co. from 1881 until 1886. The locomotive seen here was from 1885 and she would be withdrawn after seventy-four years of service during 1959.

Saturday 30 August 1952. Richard Maunsell's 'Lord Nelson' Class of four-cylinder 4-6-0s first appeared in 1926, but they proved to be disappointing performers. Improvements to the class were carried out both by Maunsell and his successor Oliver Bulleid; these included new cylinders with larger piston valves and the fitting of multiple-jet Lemaître blast pipes, all of which appeared to improve somewhat the performance of the class. Seen here standing in the yard at Eastleigh shed is No. 30851 *Sir Francis Drake*. The second of the class to be constructed, she entered service in 1928 and was withdrawn during 1961.

April 1953. During this Easter holiday, Ron spent some time travelling around and photographing the railway scene on the Isle of Wight. The images on the following twelve pages all show the ex-L&SWR Class O2 0-4-4 tank locomotives based there at that time. All had been constructed at Nine Elms Works between 1889 and 1895, and, on transfer to the island, had been given new numbers and names that reflected the island's geography. The crews must have taken great pride in their charges because locomotives always looked exceedingly clean – as can be seen in the photographs. **Friday 3 April 1953.** Seen here at Newport station, waiting to depart with the 12.20 p.m. train to Sandown, is No. 30 *Shorewell*. Transferred to the island during 1926, she had been constructed in 1892 and originally numbered 219. She would be withdrawn during 1965.

Friday 3 April 1953. No. 17 *Seaview* is parked over the ash pit at Newport shed. Constructed in 1891 and numbered 208, she arrived on the island during 1930 and would be withdrawn on the closure of steam workings in December 1966.

Friday 3 April 1953. Parked adjacent to the shed at Newport is No. 34 *Newport*. Entering service in 1891 numbered 201, she would be transferred to the island in 1947 and withdrawn during 1955.

Saturday 4 April 1953. Seen here departing from Newport station with the 11.26 a.m. Cowes to Ryde service is No. 36 *Carisbrooke*. Originally numbered 198 and entering service during 1891, she would be one of the last of the class to arrive from the mainland in 1949. She would be withdrawn from service during 1964 after seventy-three years of service.

Sunday 5 April 1953. Bearing a 71F Ryde shed code, No. 21 *Sandown* is parked in the yard at Newport shed. Constructed in 1891, she was originally numbered 205. She was transferred to the island as early as 1924, and would be withdrawn during 1966 after seventy-five years of service.

Sunday 5 April 1953. Sitting adjacent to the shed at Newport is No. 28 *Ashey*. She is bearing a 71E Newport shed code plate.

Sunday 5 April 1953. This is another view of No. 34 *Newport* sitting in the yard at her home shed, Newport.

Monday 6 April 1953. Approaching Newport station with the 9.56 a.m. Cowes to Sandown train is No. 25 *Godshill*. Another early arrival to the island's railways during 1925, she was of 1890 vintage and was originally numbered 190. She would see seventy-two years of service before being withdrawn in 1962.

Tuesday 7 April 1953. Seen here approaching Sandown station, heading the 10.28 a.m. Ryde to Ventnor, is No. 23 *Totland*. Built during 1890 and originally numbered 188, she would arrive on the island in 1925 and be withdrawn during 1955.

Tuesday 7 April 1953. Waiting for the 'right-away' from Brading station is No. 19 *Osborne*, which looks in very clean condition. Constructed in 1891 and numbered 206, she was another early arrival on the island, coming over from the mainland in 1923. She would serve sixty-four years before being withdrawn in 1955.

Tuesday 7 April 1953. At Brading station 'change for St Helens and Bembridge' we see No. 23 *Totland* again, this time heading the 1.28 p.m. Ryde to Ventnor train.

Tuesday 7 April 1953. Approaching Brading station is No. 14 *Fishbourne* at the head of a train from Bembridge.

Tuesday 7 April 1953. Waiting to depart from Bembridge station is No. 14 *Fishbourne*. An 1889-built example of the class and originally numbered 178, she arrived on the island during 1936 and would give seventy-seven years of service before being withdrawn in 1966.

Tuesday 7 April 1953. At Sandown station, No. 27 *Merstone* is seen arriving with the 1.55 p.m. working from Freshwater to Shanklin. Constructed during 1890 and numbered 184, she was transferred to the island in 1926 and would be withdrawn when steam working on the island ceased in December 1966.

Tuesday 7 April 1953. At Brading station again, No. 21 *Sandown* is seen working the 2.28 p.m. Ryde to Ventnor train.

Wednesday 8 April 1953. At Freshwater station 'for Alum Bay and Totland Bay', the fireman has just uncoupled the locomotive No. 34 *Newport* from the carriage stock after arriving with the 11.10 a.m. train from Newport.

Friday 3 July 1953. Constructed to a design by Robert Billinton and introduced during 1893, the fifty-five examples of the ex-LB&SCR Class C2 0-6-0 were all manufactured by the Vulcan Foundry. Found to be somewhat lacking in power, Earle Marsh rebuilt forty-five of them with larger boilers that proved to be of enormous benefit. The rebuilt locomotives were classified C2X and started to appear from 1908 onward. Seen here at St Leonards shed is No. 32450 looking in ex-works condition and bearing a 75B Redhill shed code. She entered service during 1894, was rebuilt in 1911 and would manage sixty-seven years in service before being withdrawn in 1961.

Friday 3 July 1953. Parked in the yard at St Leonards shed and looking rather forlorn is Class R1 0-6-0 tank No. 31335. Designed by James Stirling for the SER as Class R and introduced during 1888, the twenty-five examples of the class were primarily used on shunting duties. A large number of the class were rebuilt from 1910 on, during the tenure of Harry Wainwright, and classified R1. The locomotive seen here entered service from Ashford Works in 1888 and was rebuilt during 1915. She spent her entire working life as Hastings station pilot and would be withdrawn in 1955.

Friday 3 July 1953. Harry Wainwright-designed ex-SE&CR Class H 0-4-4 tank No. 31327 is seen here at St Leonards shed. Constructed at Ashford Works between 1904 and 1915, the sixty-six examples forming the class were powerful, free-steaming locomotives. No. 31327 was from a batch built during 1907 that would be withdrawn in 1959.

Friday 3 July 1953. Seen here in Rolvendon station yard is ex-LB&SCR Class A1X 'Terrier' 0-6-0 tank No. 32655. Constructed at Brighton Works in 1875 as a Class A1 locomotive, she would be rebuilt as a Class A1X in 1912. Originally named *Stepney* and numbered 55, she would be withdrawn after eighty-four years of service during 1960 and sold to the Bluebell Railway, where she is currently awaiting an overhaul.

Friday 3 July 1953. In ex-works condition at Ashford shed is ex-SER Class D1 4-4-0 No. 31735. Originally constructed as a Class D by Sharp, Stewart & Co. during 1901, she would be rebuilt as seen here by Beyer, Peacock & Co. in 1921. Here she bears a 75B Redhill shed code. She would give sixty years of service before being withdrawn during 1961.

Wednesday 19 August 1953. Standing in the yard on a wet day at Plymouth Friary shed is ex-L&SWR Class B4 0-4-0 tank No. 30102. Originally numbered 102 and named *Granville* by them, she was built at Nine Elms Works in 1893 and utilised for light shunting and dock duties. Note the spark arrester fitted to this example. She would be withdrawn in 1963 and subsequently sold to Butlin's for display at its Ayr camp. She is now a static exhibit at the Bressingham Steam and Gardens Museum near Diss in Norfolk.

Thursday 20 August 1953. Seen here at Exmouth Junction shed is Class E1/R 0-6-2 tank No. 32697. Originally constructed at Brighton Works as a Class E1 0-6-0 tank locomotive for the LB&SCR during 1876, numbered 105 and named *Morlaix*, she was rebuilt as an 0-6-2 tank in 1929 and would be withdrawn from service in 1959 after eighty-three years of service.

Thursday 20 August 1953. The British Railways Standard version of the LM&SR Class 3 2-6-2 tanks were numbered in the 82000 series, with forty-five examples being constructed at Swindon Works. The Southern Region was allocated twenty of the class, ten appearing during 1952 and the rest in 1954. Seen here at Exmouth Junction shed is No. 82010, which entered traffic during 1952 and was allocated to that shed. She would be withdrawn after only thirteen years in 1965.

Thursday 20 August 1953. The modern-looking lines of Richard Maunsell's design of Class N 2-6-0 were prepared during 1914, but it was 1917 before the first of this new locomotive class exited the Ashford Works of the SE&CR. A further fourteen locomotives followed from Ashford between 1920 and 1923, but it was due to post-war production of locomotive parts at the Woolwich Arsenal that a further fifty examples were constructed at Ashford using these parts together with boilers manufactured by outside suppliers. So useful were these locomotives that another fifteen were built by the Southern Railway at Ashford between 1932 and 1934, bringing the total class number to eighty examples. Seen here at Exmouth Junction shed is No. 31838, an example of the Ashford-constructed, Woolwich Arsenal-parts locomotives, which entered service during 1924 and would be withdrawn in 1964.

Thursday 20 August 1953. At Exmouth Junction shed again is ex-L&SWR Class M7 0-4-4 tank No. 30046 bearing a 72A Exmouth Junction shed code. A product of Nine Elms Works in 1905, she would be withdrawn from service during 1959.

Sunday 4 April 1954. Seen here at Nine Elms shed is an example of the NBL-constructed 'King Arthur' Class N15 4-6-0s for the Southern Railway. No. 30781 *Sir Aglovale* entered service during 1925 and would be withdrawn in 1962.

Sunday 4 April 1954. Parked adjacent to the shed at Nine Elms, bearing a 70A shed code, is ex-LB&SCR Class E4 0-6-2 tank No. 32493. Constructed at Brighton Works during 1899, numbered 493 and named *Telscombe*, she would be one of the early withdrawals of the class in 1958.

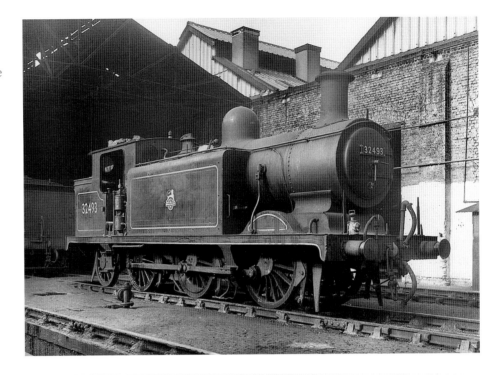

Sunday 4 April 1954. The well-proportioned locomotive seen here at Nine Elms shed is ex-L&SWR Class M7 0-4-4 tank No. 30244. Entering service during 1897 from Nine Elms Works, she would see sixty years of service before being withdrawn in 1957.

Saturday 14 April 1956. Standing on the coal road at Eastleigh shed is Robert Urie-designed ex-SR Class H15 4-6-0 No. 30475. Entering service from Eastleigh Works during 1924 and primarily intended to handle mixed traffic, she would be withdrawn in 1961.

Saturday 14 April 1956. Seen here standing in the yard at Eastleigh shed is ex-SR rebuilt 'Merchant Navy' Class 4-6-2 No. 35018 *British India Line*. Originally constructed at Eastleigh Works in 1945, she would re-enter the same works during 1955 to be the first of Oliver Bulleid's 'Pacifics' to be rebuilt in a more conventional form. Replacing the chain-driven valve gear with Walschaerts valve gear on all three cylinders and removing the air-smooth cladding, she re-entered service in February 1956, just two months prior to this photograph. She would be withdrawn in 1964.

Saturday 14 April 1956. Parked in the yard at Eastleigh shed is BR Standard Class 5 4-6-0 No. 73113. Constructed at Doncaster Works during 1955, she was one of ten locomotives of that class destined for the Southern Region to replace older locomotives being withdrawn at that time. She was allocated to Nine Elms shed and would acquire the name *Lyonesse* from withdrawn 'King Arthur' Class locomotive No. 30743. She would herself be withdrawn at the end of steam operations on the Southern Region in 1967.

Saturday 14 April 1956. Seen here at Eastleigh shed, destined to be withdrawn during the year following this photograph, ex-L&SWR Class B4 0-4-0 tank No. 30082 is looking in particularly grimy condition. Built at Nine Elms Works during 1908 and originally numbered 82, she would be withdrawn in 1957.

Saturday 14 April 1956. Bearing a 72C Yeovil shed code, ex-L&SWR Class M7 0-4-4 tank No. 30131 is seen parked in the yard at Eastleigh shed. She was one of the last ten examples of the class to be constructed at Eastleigh Works in 1911. She would be withdrawn during 1962.

Saturday 14 April 1956. Sitting in the yard at Eastleigh shed is ex-L&SWR Class T9 4-4-0 No. 30732. The graceful lines of this Dugald Drummond design are apparent here, even though his elegant chimney design has been replaced by the stovepipe variety. A product of Dübs & Co. from 1900, she would be rebuilt, as seen here, with a superheated boiler during 1927 and withdrawn from service in 1959.

Saturday 14 April 1956. Standing beside the coal stage at Eastleigh shed is diminutive ex-L&SWR Class C14 0-4-0 tank No. 30588. Originally constructed as 2-2-0 tanks, this class numbering ten examples was found wanting in its ability to handle even single coachload branch work and, by 1917, all bar three examples had been withdrawn from service. Designed by Dugald Drummond and constructed at Nine Elms Works, the example seen here entered service during 1906 and was rebuilt in 1922. Ending her days shunting at the Town Quay in Southampton, she was allocated to 71A Eastleigh at the time of this photograph and would be withdrawn the following year, in 1957.

Saturday 14 April 1956. Destined to become part of the National Collection, ex-L&SWR Class T9 4-4-0 No. 30120 is seen sitting near the coaling stage at Eastleigh shed. Constructed at Nine Elms Works in 1899, she would be rebuilt in the form seen here during 1927 and be withdrawn from service in 1963. She is currently in the care of the Bodmin & Wenford Railway.

Saturday 12 May 1956. Parked in the yard at Brighton shed is Lawson Billinton-designed ex-LB&SCR Class K 2-6-0 No. 32338. Constructed at Brighton Works and the second member of the class to enter service during 1913, she would be withdrawn in 1962.

Saturday 12 May 1956. An award-winning locomotive is seen here at Havant station. Ex-LB&SCR Class A1X 'Terrier' 0-6-0 No. 32640 was built at Brighton Works during 1878 and exhibited at the Paris Exhibition in the same year, at which it was awarded a gold medal. Originally constructed as an A1 Class, numbered 40 and named *Brighton*, she would be rebuilt as Class A1X in 1918. Sold to the Isle of Wight Railway in 1902, she would return to Southern Railway ownership at the grouping. In 1963 she would be withdrawn and sold to Butlin's to be displayed at their Pwllheli camp. During 1975 she arrived back on the island into the care of the fledgling Isle of Wight Steam Railway, where she is currently based.

Saturday 12 May 1956. Sharing duties with the locomotive in the previous photograph is 'Terrier' classmate No. 32650. Built at Brighton Works in 1876, she was originally numbered 50 and named *Whitechapel*. Rebuilt from Class A1 to Class A1X in 1920, and transferred to the Isle of Wight during 1930, she would return to the mainland in 1936 to spend many years as a Lancing Works shunter. Returned to main-line work during 1953, she would be withdrawn during 1963. She is currently under overhaul at the Spa Valley Railway.

Saturday 12 May 1956. Seen at Brighton shed on this day is ex-LB&SCR Class E4 0-6-2 tank No. 32485. Originally No. 485 and named *Ashington*, she was a product of Brighton Works during 1899 that would be withdrawn from service fifty-eight years later, in 1957.

Saturday 12 May 1956. Ex-LB&SCR Class E4 0-6-2 tank No. 32511 is also seen here at Brighton shed. This shed was allocated a large number of the class over the years and many would spend their last days allocated there prior to withdrawal. It is worth noting that twelve examples of the class would serve in France with the Railway Operating Division (ROD) during the First World War in 1918 and 1919, all being returned to the LB&SCR at the end of hostilities.

Saturday 12 May 1956. And yet another Class E4 tank at Brighton shed. No. 32468 was originally numbered 468 and named *Midhurst* by the LB&SCR. Entering service in 1898, she was one of the longest surviving members of the class, being withdrawn in 1963 after sixty-five years of service.

Saturday 12 May 1956. Ex-LB&SCR Class A1X 'Terrier' 0-6-0 tank No. 32662 is seen here at Brighton shed. Entering service during 1875 as a Class A1 locomotive, she would be rebuilt as Class A1X in 1912. Originally numbered 62 and named *Martello*, she would be withdrawn in 1963 and sold to Butlin's for display at their Ayr camp. She is currently on static display at the Bressingham Steam and Gardens Museum near Diss in Norfolk.

Saturday 12 May 1956. Brighton Works was chosen as the main design and construction facility for the BR Standard Class 4 2-6-4 tanks, and they eventually built 130 examples between 1952 and 1957. Seen here at Brighton shed during her running-in period is No. 80134; she would be allocated to the Eastern Region, but return to the Southern Region to be allocated to Bournemouth shed before withdrawal in 1967.

Sunday 14 October 1956. Still bearing the distinctive Dübs & Co. diamond-shaped builder's plate, ex-L&SWR Class 700 0-6-0 No. 30694 is parked adjacent to the shed at Nine Elms. Designed by Dugald Drummond, the thirty locomotives forming the class were all constructed in 1897. This locomotive would be rebuilt in the form seen here with a superheated boiler and stovepipe chimney during 1922, and survive until 1961 before withdrawal.

Sunday 14 October 1956. Also seen here at Nine Elms shed is ex-L&SWR Class M7 0-4-4 tank No. 30321 bearing a 70A Nine Elms shed code. Built during 1900 at Nine Elms Works, she would be withdrawn in 1962.

Sunday 14 October 1956. Another Class M7 is seen here at Nine Elms shed, No. 30241 entered service in 1899 from Nine Elms Works and would be withdrawn during 1963.

Sunday 14 October 1956. BR Standard Class 5 4-6-0 No. 73118 was delivered to the Southern Region from Doncaster Works during 1955 as part of the first batch allocated to Nine Elms shed. Seen here at that shed, she would later acquire the name *King Leodegrance* from withdrawn 'King Arthur' No. 30739. She was withdrawn in 1967.

Sunday 14 October 1956. In good clean condition, ex-SR Class 'West Country' 4-6-2 No. 34006 *Bude* is parked in the yard at Nine Elms shed. Originally numbered 21C106 when entering service in 1945, she was chosen as one of the ex-Southern Railway locomotives to take part in the 1948 Locomotive Exchange Trials organised by the then newly formed British Railways. She would give excellent performances on the London Marylebone to Manchester test runs, achieving outstanding results in comparison to the other locomotives being used. In preparation for the trials, the very long smoke deflectors were fitted. She would not go through the rebuilding process afforded to other classmates and would be withdrawn during 1967.

Sunday 14 October 1956. Seen here at Nine Elms shed is ex-SR Class 'Merchant Navy' 4-6-2 No. 35029 *Ellerman Lines*. The penultimate member of the class to enter service from Eastleigh Works in 1949, she would be rebuilt as late as 1959 only to be withdrawn during 1966. She would survive to become one of the main attractions at the National Railway Museum in York, being put on display as a sectioned locomotive.

Sunday 14 October 1956. Ex-SR Class H15 No. 30524 also at Nine Elms shed. The last of her class to be constructed at Eastleigh Works during 1924, she spent most of her working life based there. She would be withdrawn while based at Salisbury shed in 1961.

Sunday 14 October 1956. An example of the Brighton Works-built members of ex-SR Class Q1 0-6-0, No. 33003 was one of the twenty constructed there. Seen here at Bricklayers Arms shed, she would ultimately be withdrawn during 1964.

Monday 9 June 1958. The early morning summer sunshine highlights ex-L&SWR Class M7 0-4-4 tank No. 30059 at Brockenhurst station working a push-pull train from Bournemouth Central. She was an example of the 1906 batch from Nine Elms Works and would be withdrawn in 1961.

Tuesday 10 June 1958. Ex-SR Class H15 4-6-0 No. 30476 is seen here at Bournemouth shed bearing a 71A Eastleigh shed code. Another example of the class constructed at Eastleigh Works during 1924, she would be based at Eastleigh for most of her working life and withdrawn in 1961.

Saturday 7 November 1964. Seen here at Eastleigh shed is ex-SR 'West Country' Class 4-6-2 No. 34026 *Yes Tor*. Built in 1946 at Brighton Works and originally numbered 21C126, she would be rebuilt at Eastleigh Works in 1958 and withdrawn from service during 1966.

Saturday 7 November 1964. A further example of the Brighton Works-built BR Standard Class 4 2-6-4 tanks is seen here at Eastleigh shed. No. 80065 was a 1953 product of that works which, although initially allocated to the London Midland Region, would find her way back to the Southern Region to be allocated to Eastleigh shed. She would be withdrawn during 1966.

Saturday 7 November 1964. Seen here at Eastleigh shed is ex-SR Class 'Battle of Britain' 4-6-2 No. 34058 *Sir Frederick Pile*. She has had her nameplates and crests removed as she was officially withdrawn during the month prior to this photograph being taken. Constructed at Brighton Works during 1947, she had served only seventeen years in total having been rebuilt as late as 1960. Spending many years in Woodham's Barry scrapyard she is currently based at the Mid-Hants Railway awaiting restoration.

Saturday 1 May 1965. One of the earlier members of the class entering service in 1946, ex-SR Class 'West Country' 4-6-2 No. 34022 *Exmoor*, is seen here at Eastleigh shed. After being rebuilt during 1957, she would be withdrawn in the month prior to this photograph, April 1965.

Saturday 1 May 1965. The third member of the class to be constructed during 1941, ex-SR Class 'Merchant Navy' 4-6-2 No. 35003 *Royal Mail* is seen at Eastleigh shed. Rebuilt as seen here during 1959, she would be based at Exmouth Junction shed 72A for most of her working life before withdrawal in 1967.

Ron managed to visit the Isle of Wight again during the August bank holiday in 1966, this being only months before the closure of the island's remaining steam railways in December that year. Efforts to keep the locomotives clean had reduced somewhat, but they were all still identifiable. The following seven pages show the ex-L&SWR Class O2 0-4-4 tanks at work on **Saturday 27 August 1966**. No. 27 *Merstone* is seen at Shanklin station waiting to depart with the 9.30 a.m. train to Ryde Pier.

No. 24 *Calbourne* is approaching Shanklin station with the 9.00 a.m. departure from Ryde. Originally numbered 209 and, having entered service in 1891, she would be transferred to the island in 1925. Withdrawn at the end of steam workings on the island, she would be sold to the Isle of Wight Steam Railway where she is now based.

The driver of No. 22 *Brading* is seen 'adjusting' the Westinghouse pump prior to departure from Shanklin station. Built during 1892 and originally numbered 215, she would reach the island in 1924 and work until the closure of the island's railways in December 1966.

No. 35 *Freshwater* is waiting to depart from Shanklin station. Entering service in 1890 and numbered 181 by the L&SWR, she would reach the island in 1949 and be withdrawn towards the end of 1966.

No. 31 *Chale* is waiting to depart from Brading station at the head of the 10.44 a.m. Shanklin to Ryde Pier train. Built in 1890, originally numbered 180 and transferred to the island during 1927, she would also finally be withdrawn from service in March 1967 having given seventy-seven years of service.

On the approach to Brading station, the driver of No. 20 *Shanklin* is preparing to exchange the single line token while working the 10.40 a.m. Ryde Pier to Shanklin train. Originally numbered 211 when entering service in 1892, No. 20 would be one of the early transfers of the class to the island during 1923. She was withdrawn at the end of 1966.

Seen here departing from Brading station with the 11.10 a.m. Ryde Pier to Shanklin service is No. 22 *Brading*.

Waiting for the 'right-away' at Brading station is No. 28 *Ashey*. This photograph clearly shows the extended coal bunkers fitted to the Class O2 locomotives working on the island.

Seen approaching Brading station is No. 31 *Chale* at the head of the 11.40 a.m. Ryde Pier to Shanklin service.

On the approach to Brading station, No. 33 *Bembridge* is seen with the 12 noon Ryde Pier to Shanklin train. Built in 1892 and originally numbered 218, she was transferred to the island in 1936 and would be withdrawn at the end of 1966.

Standing at the Pier station in Ryde is No. 35 *Freshwater*.